16 TOPICS ON HOW YOU CAN HELP THE PLANET IN YOUR EVERYDAY LIFE

A SUSTAINABLE CLIMATE CHANGE-FRIENDLY BOOK

TEYA FRANCE

16 WONDERFUL TOPICS ON HOW YOU CAN HELP THE PLANET IN YOUR EVERYDAY LIFE

A Sustainable Climate Change-Friendly Book

I'm offering this simple handbook to you, full of exceptional ways to make many differences to improve various things on the planet, including Climate Change situations for anyone in the world who wants a reference book of suggestions for anything that can be done to help, regarding all kinds of efforts that many people are trying to make in many places everywhere.

After two years of studying how people can deal with Climate Change and some other issues on the planet as well, such as excess plastic, I found that there are essentially sixteen topics that people might want to cope with this basically in one handbook. 70% of Climate Change emissions are actually issues that are really dealt with by human beings in their everyday life, hence the title, "Everyday Life."

The first topic, "Minimizing Electricity & Household Hints" specifically refers to how <u>most of our climate change emissions are produced through the creation of electricity</u>. Electricity is created by power plants, and vast emissions rise into the sky as electricity is formed and travels to cities and homes everywhere. This electricity is the product that we are all actually responsible for putting out from our homes and businesses, for climate change. This first section discusses how we can put out less electricity from our homes.

I decided to make this handbook of suggestions really easy to follow. The suggestions don't give instructions about what Climate Change is – instead, you just find simple, easy, step by step ways to do things that help. It's easy to go on to the next topic and discover the suggestions of the next topic of How You Can Help The Planet.

The three topics that stand out at the end are "13. The Stormy Times of Climate Change," "14. Dry Weather & Heat Waves," and "15. Forest Fires." These sections give professional preparatory and safety instructions for Climate Change weather conditions. We hope you don't need them, but if you do, we hope they help.

Gotham Books

30 N Gould St.
Ste. 20820, Sheridan, WY 82801
https://gothambooksinc.com/

Phone: 1 (307) 464-7800

Published by Gotham Books (March 30, 2023)

ISBN: 979-8-88775-226-6 (sc)
ISBN: 979-8-88775-227-3 (e)

Because of the dynamic nature of the Internet, any web addresses or links contained in this book may have changed since publication and may no longer be valid.

The views expressed in this work are solely those of the author and do not necessarily reflect the views of the publisher, and the publisher hereby disclaims any responsibility for them.

The information for this handbook all comes from the Internet, and suggestions for further reading are at the end of each section. Pictures come free from Unsplash, Pixabay, and 'free to share' from Edge and WaveBrowser, as well as what was purchased from iStock.

All proceeds from the sale of this book will be used to fund environmental projects to help the world.

*CLIMATE CHANGE IS CAUSED BY
EMISSIONS OF CO2 AND OTHER GASES
RISING UP FROM SOCIETY'S WAY OF LIFE
INTO THE AIR.*

*THIS HANDBOOK GIVES TIPS
ON WHAT YOU CAN DO
IN YOUR EVERYDAY LIFE
TO HELP END CLIMATE CHANGE.*

TABLE OF CONTENTS

1. MINIMIZING ELECTRICITY & HOUSEHOLD HINTS

Most of our emissions are produced in the creation of electricity – if we reduce our electricity usage, we help end climate change.

A Single Light Bulb

A single new style light bulb can save between $40 and $135 in electricity costs over its lifetime. If every U.S. household replaced <u>one</u> regular light bulb with a CFL, it would eliminate the same amount of air pollution as taking 7.5 million cars off the road.

Five Lights

Replace your five most frequently used lightbulbs with newer kinds, and you will help the environment while saving up to $200 a year on energy bills. This kind of lighting lasts 10-15 times longer.

Turning Off Lights

Save electricity by turning off lights when leaving a room.

Unplug Things at Night

It's good to <u>unplug</u> these devices: the TV, DVD player, stereo, computer and printer. About 7% of household electricity is used to keep devices warm at night.

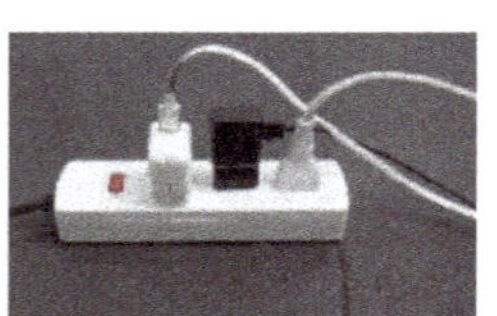

You can get a surge protector strip of plugs with an on/off switch to make it easy to turn off several devices at a time.

It's a very good idea to disconnect the energy of your <u>office equipment</u> as well – computers, printers, fax machines, etc. – when they are not going to be used!

Heating and Air Conditioning

Heating and cooling accounts for almost half of your energy bill.

☐ You can turn down the heat while you're sleeping at night or away during the day.

☐ Consider setting up room-by-room heating and air conditioning or use portable equipment.

☐ Try setting your thermostat at about 70F in winter and 60F in summer to save energy.

☐ If your heating and cooling equipment is old, you could replace it with Energy Star-certified equipment.

☐ In heat waves, you need to take care of yourself – plan for them and get portable cooling equipment in advance. Make sure you're prepared so you're going to be in good shape!

☐ If it's not incredibly hot, try to use a fan instead of an air conditioner. The air conditioner uses a lot more power and energy than the fan.

☐ There are many kinds of very affordable small 'air conditioners' that use humidity for cooling. Get one with a big water tank, and know that these units are often very small.

☐ Try looking at heat pumps/air conditioners – they don't burn fuel for energy, they get it from the outdoors. There are huge savings and rebates on these.

☐ See if you can dress for the weather, and adjust layers before adjusting the thermostat – use a blanket if you're cold.

☐ Use the power of the sun – open the blinds and turn off the lights.

☐ Open windows to help ventilate your home, office, or car.

Other Household Suggestions

☐ If you see any products that are marked 'green,' these are very good ones to support by purchasing them. They also often come with tax breaks.

☐ Buy items in the store with the least packaging.

☐ Purchase paper made from hemp, sugar cane, seeds, or even ground stone, instead of using paper made from trees.

☐ Store all solvents in airtight containers in order to avoid putting gases in the air, and try to use water-based paints, not oil.

☐ Use rechargeable batteries with a charger. Billions of batteries are bought and disposed of every year.

☐ Burning garbage adds smoke to the air – that's air pollution.

☐ Cigarette smoking adds to air pollution, especially indoors, so smoke outdoors or try to quit.

☐ Use a propane or natural gas grill instead of charcoal.

☐ Use an EPA-certified wood stove or fireplace instead of oil.

☐ It's a very good idea not to heat your home with a gas stove.

☐ Have your gas appliances and heater regularly inspected and maintained.

☐ You can put filters on your chimney – harmful gases are often emitted that affect the air quality.

Major Household Appliances

On the average, home appliances – including clothes washers, dryers, dishwashers, refrigerators, freezers, air purifiers and humidifiers – account for 20 percent of your home's total electrical bill.

Energy Star Rating

Energy Star is a U.S. and Canadian government rating for appliances that will significantly reduce your electrical usage. An Energy Star-certified appliance will use anywhere from 10 to 50 percent less energy than a non-energy-efficient equivalent.

Energy-efficient appliances, such as those with the Energy Star rating, are often more expensive, but you usually get incentives, and the energy savings you'll be getting from them usually makes up for it.

Rebates or Incentives

Utility companies can often help with thousands of dollars on the new energy-efficient appliances, so people sometimes save more on the new ones than they spend on them. Sometimes you can also get a rebate when you return an old appliance!

Water Heaters

The standard residential water heater is the highest energy consuming appliance in our home, after household heating and air conditioning.

Refrigerators

Energy Star qualified Refrigerators use up to 40 percent less energy than the conventional models that were sold in 2001. An older model uses lots of energy in many North American homes.

Clothes Washers

An Energy Star-certified clothes washer uses about 50 percent less energy and water than standard washers, at a time when many areas are trying to economize on water. When you replace your conventional washing machine with an energy-efficient model, you can expect to save up to $50 per year on utility and water bills.

Look for the new machines that have <u>microfibre filters</u> built in. These help keep small fibers of polyester and nylon from being sent into the ocean, and the front-loading models are best for this. Front-loading clothes washers also require less laundry detergent than top-loading washers, so we save on laundry soap with them, as well.

Clothes Dryers

A typical dryer uses more energy than a new energy-efficient refrigerator, washing machine, and dishwasher combined. Energy Star-certified dryers use 20 percent less electricity than conventional models, which will save you over $200 in electric bills over your dryer's lifetime.

Of course, if you can dry your clothes on a laundry line, it's a very good way to save electricity.

Dishwashers

Dishwashers may not use as much power as a constantly running refrigerator or a high-heat clothes dryer, but the electricity and water that is needed to run a dishwasher cycle adds up. Installing an energy-efficient dishwasher will save you around $25 a year in electricity costs.

Some Articles On The Internet That Have Further Information:

Enercare, "14 Energy Saving Tips to Boost Efficiency at Home," 2023

Direct Energy, "25 Energy-efficient Tips To Lower Electricity Costs," 2023

Inspire Clean Energy, "How to Save Energy at Home: 26+ Best Ways to Save Energy Today," 2023

2. YOUR OWN SOLAR ENERGY SYSTEM

In order to be free of any responsibility of emissions from your household for electricity, you can, in a very affordable way, get solar panels installed on your rooftop.

Call An Energy Contractor For Your Area

Get some advice on what's available in your area. They should be able to give you an estimate for both the preparatory renovations, and then arrange installation of a no-emission energy system for your house. They should know the local building codes and local utility companies in your area, and any red tape you might have to deal with.

How Easy Is It To Afford Your Own System?

Installing a solar energy system will immediately eliminate all the monthly energy expenses of your residence. If you get it with a loan, you can make the loan payments with the same monthly amounts you pay now, until the loan is paid off. At that time, your energy will no longer cost you anything.

A solar energy system is likely to cost from $14,000 to $36,000 USD, depending on the size of the house. The U.S. government gives a 30% tax credit discount on the whole project. Canadian homeowners can receive up to $5,000 for the preparatory energy-efficient retrofit, and the solar energy system can qualify them for a rebate of up to another $5,000.

Some of the provinces and states offer very good incentives, and often there are some on a local level as well. You can find the information on the Internet.

There are significant loans being given by companies and banks, that expect a payback in 7 to 20 years. They often give very good rates for these energy upgrade loans, and Canadians can likely get a very good interest-free loan from the Canada Mortgage and Housing Corporation (CMHC).

You Can Have A Back-Up In Case You Run Short On Energy

You usually have the option of being 'on-grid' to your big, local company, as a back-up to your own system, and on sunny days it stores excess energy created by your system. The company meters what comes in and out, so at the end of the accounting period you pay for the net energy received, minus the net amount of energy you have given. Being on-grid is called Net-Metering in Canada.

Preparing For A Solar Panel System Installation

This preparatory renovation is a significant part of getting the energy system. These are the types of things that need to be done, especially to make sure energy from your new system is efficiently used, by:

☐ Making sure your roof is in good shape.

☐ Airtightness and Insulation, so warm or cool air doesn't escape through cracks around windows and doors. Insulation in your attic, and around your water heater.

☐ Updating heating, cooling, and ventilation. Consider room-by-room heating and cooling to be more efficient. Replace heating oil system.

☐ Major appliances and lighting. Looking at getting an energy-efficient refrigerator and replacing other outdated appliances and lighting.

☐ Other household upgrades. This might be the perfect time to get a new kitchen counter, or new carpeting!

☐ **NOW IT'S TIME TO GET YOUR NEW ENERGY SYSTEM INSTALLED!**

Some Articles On The Internet That Have Further Information:

Smart Reno, "Solar Panel Guide for Homeowners in Canada," 6/2022

Products, "A Guide To Energy Retrofits In Existing Homes," 7/2022

3. BEING AWARE OF WASTEFULNESS

Wastefulness is very high in North American society – the creation of new manufactured goods generates about 45% of global emissions, and many items are thrown away very soon when they could be resold.

Shopping

The average product causes emissions of <u>more than six times its own weight</u>. Much of what we buy on shopping trips ends up in the landfills soon afterwards.

Clothes According to Fashion

With about six microseasons a year, there are always brand new clothes and new styles to purchase year round.

With the situation of nylon and polyester microfibers of plastic going into the ocean from our laundry water (see the chapter on Plastics), older clothes are less likely to emit these fibers. It's a good reason to purchase new clothing less frequently. Give used clothing to thrift shops or consignment stores - it shouldn't be thrown into the garbage.

Christmas Could Be Changing!

We could consider changing our traditions rather than exchanging presents.
Purchasing gifts for people who aren't close to us could be causing a lot of

unnecessary
manufacturing.
Also, people might
not need what we
give them and they
end up in the land
fill! It is known
that this happens
because a lot of
newly purchased
products, still in
their original
packaging, appear
in the landfill right after Christmas.

Packaging

With online shopping being the preferred purchasing method for many people,
we are generating immense waste in unnecessary packaging. If you can get all
your items delivered together instead of in several boxes, it would keep the
number of packages down for the sake of trees and plastic usage.

You might find packaging of all kinds
may be much less necessary than we used
to think – for instance, you could get
bulk foods or vegetables in the grocery
store by bringing your own bags and
containers, instead of using the plastic
bags supplied by the store!

Use Online Shopping Networks, Thrift Stores, & Bartering

You can find alternate ways to sell and purchase what you need. Facebook has
'Marketplace,' where all kinds of used and almost new items are bought and
sold, many of them in good condition and for a very low price, or even for free.
There are also 'Buy And Sell' papers and online, and 'Craigslist' online, so people
can find it easy to sell or give things away which still have value.

You can look into <u>bartering</u>. There are some large and exciting barter networks where people post items, offer services, and purchase things using points systems or other barter company methods. They can be found in most metropolitan areas.

If you can find ways to give your cast off clothing and other items to <u>thrift stores, barter networks, secondhand clothing stores</u>, and <u>consignment shops</u>, and think of patronizing these establishments before heading to the mall or other places to find brand new things, this would help your budget, and help climate change by reducing waste.

Food Waste

Approximately 30% of food is wasted – and it has been winding up in landfills! A lot of electricity is used in the preparation and handling of the food that is wasted. Also, methane emissions are produced in the landfill from this food,

and these are much worse than CO2. Landfills are the third largest methane source in North America.

Some governments have arranged for recycling companies to collect people's food scraps and turn them into fertilizer for farmers for their fields. Many municipalities now provide <u>small buckets</u> to people in their communities for collecting food scraps.

The "Best Before Date" On Grocery Items Is NOT an Expiration Date

Packaged food is usually still good, but is thrown away by the store or the consumer on the "best before date." This is only the date when the unopened package of food is the freshest. The food is still good and edible for a week or ten days afterwards. Don't throw the cartons of food away on that date.

Here are some ways to prevent food wastage:

☐ Food spoilage at home occurs due to improper storage, lack of visibility, and misjudged food needs. Try and eat food while it is still good.

☐ Don't forget to eat leftovers.

☐ Make meal plans and stick to shopping lists so you don't buy too much.

☐ Realize that unplanned restaurant visits or food delivery can lead to food going bad at home before it can be used.

☐ Buy herbs and vegetables in the amounts that you need them in.

If food waste were eliminated from landfills, it would make a difference to emissions like removing one-fifth of all the cars from the road.

Grocery Stores & Restaurants Could Give Away Excess Food

Food that has reached its "Best Before" date could be donated to Food Banks and projects for people with low incomes, senior citizen homes or nursing homes.

Some Articles On The Internet That Have Further Information:

Almost Zero Waste, "90 Zero Waste Tips For Beginners (Impactful & Easy Habits)," 2023

Sustainable America, "A Zero-Waste Christmas Guide For A More Sustainable Holiday," 12/2021

The Modest Wallet, "21 Sites Like Craigslist to Buy or Sell Used Stuff" 6/2021

4. COWS & THE CLIMATE

Believe it or not, there are about 1.5 billion cows in the world – about one cow for every five people – compare that to 1.3 billion cars on the road.

Cows have four stomachs, and one contains bacteria to help digest its food. The bacteria emits a lot of methane gas, so every cow emits 400-600 liters of methane per day. If a water bottle for drinking holds one liter, you can imagine how much methane cows are putting into the air.

Methane is about 80 times worse for climate change as CO2, so this situation with cows is very serious. The amount of emissions that are put into our atmosphere by cows is 14-18% of the global emissions!

To reduce this problem, we need to eat less beef to reduce the demand for it, and get the cattle industry to reproduce fewer calves. They have been feeding other types of food to the cows to lower the methane, and it has been a somewhat successful, but they might need to revise their industry to make a difference. Dairy cows are still needed.

It's recommended that all of us lower our beef consumption and eat about <u>one</u> hamburger per week!

—

Some Articles On The Internet That Have Further Information:

World Resources Institute, "6 Pressing Questions About Beef and Climate Change, Answered," 3/2022

Forbes, "The Potential For More Climate-Friendly Cows – Unnecessarily Delayed," 11/2022

5. IMPROVEMENTS IN YOUR VITAMINS & MINERALS

Restoring Health To Our Soil

After fifty years of industrial farming methods, awareness of the health of the soil has grown, while most of the <u>meat</u>, <u>dairy</u>, <u>eggs</u>, <u>fruits</u>, and <u>vegetables</u> available in supermarkets today are still produced using these methods.

'Regenerative agriculture' is the term for a new style of farming that restores health to the soil. It focuses on topsoil regeneration, and increases the number of good insects and wildlife. Grazing animals help the land through pasture rotation, and a diverse mixture of crops are grown. With improved soil, food that has more nutrients can be grown for many people.

This method of agriculture restores the fresh water that helps farms - natural fertilizers are used which don't leave a chemical residue in the water supply.

If you buy organic fruits and vegetables, you will help encourage farmers grow this kind of produce.

It's good to visit your local farmers markets, and buy organic fruit and vegetables in the stores!

Vitamins Are Very Good To Have Right Now

Because the soil our food is grown in has supported many crops, it is a bit overused and the food value is about half of what it used to be. Therefore, it's very good for people to take vitamins until we have better nutrition in our soil.

If we're not getting the vitamins or minerals that we need, we could have trouble with our thoughts, moods or even depression. It's also good to eat a variety of foods and not the same thing every day.

Recommended Vitamins

☐ Every person should be taking a ***multivitamin*** every day. Children should have child-size multivitamins, and it is best for both adults and children to always take a quality brand – the fun and colorful pills might not supply all that we need.

☐ ***Vitamin C*** comes from oranges, and is important for all of us in several ways - it gives us a feeling of sunshine in our consciousness. It's okay to take two per day if you feel like it!

☐ ***Iron*** is important for women under fifty-five. Anyone can also take a low dosage tablet every few days.

☐ ***If you don't eat much meat and fish,*** it's very important to take ***B-Complex*** and

Calcium/Magnesium, as well. Without these, you could be susceptible to feelings of moodiness.

☐ Most of the other vitamins that are required are provided in the multivitamins.

Preparing Your Daily Vitamins

If you take more than a couple of vitamins every day, you can put the tablets into daily containers. If you shop online, just look for 'vitamin boxes.'

—

Some Articles On The Internet That Have Further Information:

Harvard Health Publishing, "Listing of vitamins," 8/2020

Amy B. Scher, "Muscle Testing: Getting Answers From The Subconscious Mind," 2021

6. PLASTICS & YOUR LAUNDRY WATER

As of 2017, the world has produced <u>8.3 billion tons of plastic</u>, but only about nine percent of plastic has been recycled. Unfortunately, about 25% of the emissions in the world are due to the plastics industry.

When you do your laundry, tiny fibers of polyester and nylon plastic come off your clothes into the water. A regular washload could actually have around

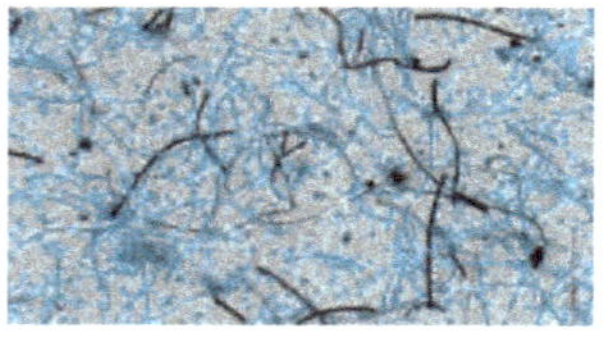

600,000 fibers in it. Fleece especially emits a lot of microfibres. They go into your laundry water and end up in the ocean. There are so many in the ocean, you cannot go anywhere at all without finding these fibers in the water. Also, the amount of fragments that have sunk to the sea floor is far greater than the number floating around in the ocean.

Ways to reduce the microfibers in your laundry wash water

☐ Get a microfiber filter for your washing machine, or buy a new one with a filter built in.

☐ Use cold water – it reduces shedding by about 50%.

☐ You could wear your clothes a day or two longer, so you can have washloads less often.

☐ Wash full loads so you don't use extra water.

☐ Keep your clothes longer so they get older – older clothes shed less.

☐ Why not contribute to a carbon offset project to compensate for what your laundry water is sharing with the ocean!

☐ Use front-loading washers, if possible - they produce fewer microfibers than top-loading.

Ways To Avoid Using Plastic In General

Here is a list of methods to avoid using plastic. We know you can come up with many, many more!

☐ Bring your own reusable bag when you go shopping – keep one or two in your purse, car, or bike carrier.

☐ Avoid getting convenience foods packaged in plastic if you can help it .

☐ Re-think your food storage – instead of sandwich bags, you can use a jar, glass container, reusable lunch bag or beeswax food wrap.

☐ Buy laundry detergent in boxes, avoid liquid in plastic containers.

☐ Paper bags are better to use than plastic ones.

☐ Cloth napkins feel nice and they reduce the number of plastic trash bags required to dispose of the paper ones.

☐ Switch to shampoo bars, sometimes sold in shampoo areas of drugstores and grocery stores.

☐ Re-think your deodorant that comes in a plastic container.

☐ Sometimes there's no way around using a plastic item!

☐ Bring your Styrofoam coffee cup home from the coffee shop, wash it and put it in your cupboard for another time. It can be reused once or twice.

☐ 1 billion toothbrushes are thrown out every year in North America – try a wooden one next time from a health food store.

☐ It's very good to keep Styrofoam for packaging since it's hard to get it recycled and you don't want to send it to the landfill.

☐ Use dry toothpaste instead of taking it out of a plastic tube. (50,51)

☐ In order to stop buying single-drink bottles and throwing the bottles away after one use – you could get a Sodastream, Drinkmate OmniFizz, or other Water Carbonizer – many people find they love the bubbly, carbonized water, and you can use flavour enhancers and soft drink flavors in them. Even though they are initially pricey, they are well worth it because they replace the need for single-drink bottles and cans.

☐ Use Bamboo utensils instead of plastic forks and knives.

☐ Make your own cleaning products out of simple, even food-grade, ingredients & store them in refillable glass spray bottles.

☐ Use Dishwashing blocks instead of liquid soap in plastic bottles.

☐ Use Sustainable sponges – conventional sponges are often chemically dyed, emit microplastics and are not biodegradable.

☐ For scouring dishes and pots, there are coconut, hemp or bamboo bristle brushes.

☐ Build a zero-waste kit for bulk shopping – this is inexpensive and it reduces packaging waste. You'll need some eco-friendly bags and glassware for shopping and storage.

☐ Organic cotton produce bags can be used in place of plastic bags when shopping for fruits and vegetables.

—

Some Articles On The Internet That Have Further Information:
Earth911, "How To Reduce Your Exposure To Microplastics," 5/2022
My Plastic Free Life, "100 Steps to a Plastic-Free Life," 2007-2023

7. SUSTAINABLE COMMUNITIES

What Are Sustainable Communities?

These are eco-villages, residential housing developments, multi-unit residential buildings, or grassroots community movements where the members choose very ecological standards in their lives.

They usually depend on solar energy, have very good insulation and heat recovery systems.

Some of them incorporate several features that integrate the community closely. Community gardens and orchards, edible landscaping, collective rainwater harvesting, and district energy sharing systems distinguish these residents as holistic thinkers about what sustainability means.

According to The Global Ecovillage Network, about 420 sustainable communities exist around the world, 150 are spread around the United States, and somewhere around 70 are across Canada.

—

Some Articles On The Internet That Have Further Information:

Green Building Canada, "Sustainable Communities in Canada," 10/2020

Microsoft Bing, "Sustainable communities in the world images," 2023

8. TRANSPORTATION

Electric Vehicles

If you're ready for a new car, consider getting an energy-efficient one such as an electric vehicle. Some come in hybrid models, which run for 40 miles or so on battery, and then they switch over to gasoline. The others run all the time on battery. There are new models all the time.

You can charge them up at home, and it's best if you have a solar energy system in your house to do that with, so you are giving your car clean electricity to run on. They can also be topped up from the public car charger system.

A Bicycle Lifestyle

A no-emission mode of transportation! If you can work at home, you could enhance your lifestyle with a lot more bike riding!

☐ It's very easy to do local errands because you can park your bike right outside where you need to be.

☐ You can shop less and more often so you only have to carry supplies for two to three days in advance. This allows you to have fresher food.

☐ If you need to, you can order heavy items such as cartons of juice and canned food in bulk every few weeks and have them delivered.

☐ If you do your shopping from your bike, there are various methods of carrying your purchases – a backpack, a basket on the front or rear, panniers over the rear wheel, or you can get a trailer.

☐ Try to ride on alternate traffic-free routes to avoid breathing fumes from cars.

In the U.K., there is the popular "National Cycle Network." This is 5,000 miles of traffic-free cycling routes in Scotland, England, Wales and Ireland that you can see on the maps listed in the reference at the end of this chapter.

Driving Smarter

In order to minimalize your driving and keep from putting too many gas emissions into the air, it's a good idea to:

☐ consolidate errands into a single trip

☐ shop locally if possible

☐ keep the weight within the car as light as possible

☐ drive less, especially on days with smog in the air

☐ accelerate gradually

☐ drive within the speed limit

☐ get regular tuneups because that will ensure the car is efficient on gas

☐ keep your tires properly inflated so you don't burn excess gas that way

☐ If you have a Smog Check program in your area, try and support it

☐ take your lunch to work to avoid going out in your car at lunchtime

☐ it's really good if you can work from home – using online communication is very convenient, instead of traveling back and forth in your car

Public Transportation

This is a very highly recommended way of going places, and it saves on gas even if there are just seven people on a bus.

You can save thousands of dollars a year by taking public transit instead of driving. The amount of emissions created is much, much less per person and per city. You should be able to find information on your computer or mobile phone that shows transit information and schedules for your area.

A Car – Only When You Need It!

Carsharing is a way for people who don't own a car, to use a car when they need one, for an errand or appointment.

You can phone or reserve a vehicle online that is parked in your neighborhood, and walk over to it when you're ready to get it. You can drive to where you need to go, and leave it in that area or back in your neighborhood. The major carsharing companies in North America are: *Getaround, Turo, CarSharing, The Hertz Corporation, Lyft,* and *Zipcar.* The membership fee usually includes gas, insurance, and maintenance of the vehicles.

Carpooling

This is a way to save money, gasoline and cause fewer emissions on regular routes.

If a few friendly commuters ride together in a carpool 20 days a month, it reduces the air pollution emitted, the driving costs get  lowered by at least 40-50 percent, and wear and tear on the vehicle is reduced.

There are many carpool websites available on the Internet for information in your area.

Air Travel

We all love to jump on a plane to go to a meeting or on vacation. Now we need to look at things differently.

Air travel puts out a lot of emissions high in the atmosphere, and these are especially problematic. Air travel is responsible for about 3% of global emissions, but it is predicted to be the greatest air polluter in the future if we keep flying a lot! It would make a great deal of sense to do business and have social meetings over the Internet to avoid air travel. Many people are not using air travel at all and they now plan to take their vacations close to home.

Almost 4 billion passengers were carried by air flights around the world in 2022 – about half the world's population!

Long distance flights, without any stops, such as from the west coast to the east coast of North America, are more efficient on fuel than shorter ones, so these are more worthwhile. Shorter trips are better made on land by car, bus, or rail.

The heavier an aircraft is, the more emissions it creates. Cargo and baggage make up a large portion of an aircraft's weight, so each individual traveler can contribute by packing lightly - even one less pair of shoes from all passengers would make a noticeable difference to the emissions.

All airlines are looking at what they can do to help the emissions situation, and some of them help passengers do carbon offsetting to make up for their share of emissions that come from flying. These airlines help with carbon offsetting for their customers: 1. The Emirates 2. Qantas 3. Virgin Australia 4. Delta Airlines 5. British Airways 6. Jetstar 7. Air New.

Some aircraft are now being built with graphene, which is very light and strong, and this will undoubtedly revolutionize the aviation industry before long.

—

Some Articles On The Internet That Have Further Information:
Joro, "How to Fly Sustainably: 6 Tips to Cut Your Carbon "Flight" Print," 10/2022

Wired, "The 17 Best EVs Coming in 2023," 2023

Sustrans, "The National Cycle Network," excellent website of the U.K. Network

9. HOW TO 'ECO' YOUR COMPANY

☐ Go paperless or use less paper. Use email whenever possible. Do away with paper billing.

☐ Use natural light during the day.

☐ Your energy costs will go down if you're using the new LED lightbulbs.

☐ Choose green suppliers and partners.

☐ Provide separate recycling bins for paper and plastic.

☐ Use recycled materials for packaging.

☐ Unplug, or turn off a power bar at night, that has equipment on it that stays warm when not being used. This uses 7 percent of the energy of your office.

☐ Check with your web host to see if they're using no-emission sources.

☐ Commercial solar panel installation will save you money as it will end your electric bill (See the Chapter on "Your Own Solar Energy System").

☐ Encourage employees to carpool, walk, or bike to work.

☐ If your employees can work from home whenever possible, office energy use is decreased, and vehicle pollution is greatly decreased.

Some Articles On The Internet That Have Further Information:

EcoMENA, "Ways to Make Your Business Eco-Friendly," 3/2022

Bestpackers, "How to Make Your Company More Eco-Friendly," 4/2022

10. TREE PLANTING

☐ ***Plant 3-5 Trees A Year For Twenty Years!*** Trees absorb CO2, so if every person on the planet did this, it would possibly negate about half the emissions emitted by society since 1960.

☐ Many people and companies feel that trees are a renewable resource, and they are not when it comes to climate change! The green leaves absorb CO2 emissions from the atmosphere through photosynthesis. Because of this, trees need to stay and grow for a long time, and not be considered a renewable resource that gets planted and cut down soon after.

☐ Get these products made out of sugar cane, ground seeds, ground stone, hemp, or bamboo instead of using regular products made from wood:

- printer paper
- paper towels
- cardboard
- Christmas wrapping paper

☐ You can grow 'tiny forests' in local empty lots in your city.

☐ Why not go tree planting near your town with your family or friends for an afternoon!

☐ Some people give up their full time job to do tree planting instead.

☐ Millions and billions of trees are being planted by some countries because they know that forests absorb a lot of CO2.

☐ However, three times as many trees are still being cut down as are planted. The trees that are cut down are used for paper and construction. Try not to cause wood to be cut down and used – it's needed to be in a tree.

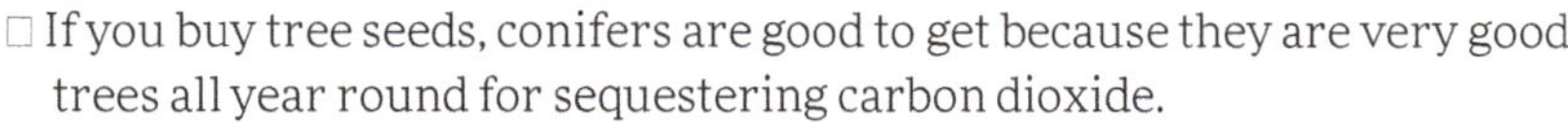

☐ You can grow seedlings for tree planting in your kitchen, getting seeds and seed starter trays from the store.

☐ If you buy tree seeds, conifers are good to get because they are very good trees all year round for sequestering carbon dioxide.

Some Articles On The Internet That Have Further Information:

8 Billion Trees, "How Many Trees Are Planted Each Year? Full List By Country, Type, Year," 9/2022

Woodland Trust, "How To Grow A Tree From Seed"

11. AMAZING HOUSES WITHOUT WOOD!

Since so many trees are getting planted, it would be wonderful to see more homes that are built without cutting any trees down! Here are four beautiful types of homes that can be made without wood:

Steel Beam Houses – like wood frame houses, but made with steel beams – with many more options than wood, house kits are available on the market.

Versatile 3D Printed Houses – a new process, becoming very popular, that uses the technology of a 3D printer on a track, software controlled, building houses layer by layer. Parts of the house are joined together.

Benefits are:

- Very good detail
- Cost effective
- Time efficient
- Reduced complexity
- Easy duplication

Houses can be built with a variety of materials, and last for 170 years.

Concrete and Cement Houses - Very strong, versatile and durable, with comfortable interiors away from hot or cold weather.

Materials can be found close at hand, and can

include recycled byproducts. The house can be recycled for other purposes at the end of its useful life.

Graphene Dome Houses – very affordable modular kits come from countries such as China, very strong in earthquakes and storms due to the graphene and geodesic dome style. House kits come with solar.

—

Some Articles On The Internet That Have Further Information:
Google, "Steel Beam House Kits Images"
GreenMatch, "Top 13 Alternative Housing Ideas," 2021

12. HOW TO HELP WILDLIFE

The love of plants and animals is felt differently by every culture.

☐ Take a walk outdoors and appreciate the birds and wildlife.

☐ Volunteer at a local wildlife refuge and at conservation projects.

☐ Watch wildlife movies to learn more about them.

☐ Sign petitions that protect wildlife, and 'share' on social media.

☐ Try to reduce your environmental footprint.

☐ Never buy ivory of any kind!

☐ Teach your children about wildlife by visiting a wildlife refuge, nature reserve, wildlife hospital, zoo, or national park.

☐ Donate money to wildlife non-profit organizations.

☐ Drive more slowly so you can brake for animals.

☐ Avoid using herbicides, pesticides, or any toxins on your property.

☐ Pick up litter that you see on the ground – it could save the life of an animal!

☐ Adopt an at-risk animal species, or adopt an acre. Organizations such as WWF, Defenders of Wildlife, and National Wildlife Federation do this.

☐ Protect, restore, and manage the land and habitat on your property. You can work with professionals like county foresters and wildlife biologists.

☐ Learn how to build a Monarch Butterfly waystation to help conserve and protect them.

☐ Track bird species for the National Audubon Society in their "great backyard bird count" across the U.S. and Canada.

☐ Donate to land trusts in your area that conserve important wildlife habitats for future generations.

☐ Take classes and workshops to learn more about wildlife conservation.

☐ Consider a career in wildlife conservation!

☐ Join Associations of wildlife enthusiasts.

☐ Consider your impact on the local habitat when buying a house.

☐ National Parks play a key role in preserving wildlife habitats. They maintain healthy ecosystems, clean air and water for wildlife. They try and teach us all environmental stewardship.

☐ You can join a Streamkeeper group in your area to open a stream or river so that fish can swim up it and lay their eggs. Sometimes these waterways get blocked off from urban construction. Fish dwindle in number and the species that eat them don't get enough to eat.

—

Some Articles On The Internet That Have Further Information:
wikiHow, "How to Legally Protect Wildlife," 7/2020
NRDC, "How Can I Protect Wildlife in My Community?" 6/2018

13. THE STORMY TIMES OF CLIMATE CHANGE

When water falls it is very wise to save it, or conserve it for future use.

If you live in a rainy area, you and your neighbors can limit your usage of water so you can share it with areas far away that will likely be experiencing drought in the foreseeable future. Your government could make plans ahead of time for these areas to come and get water at the times they need it.

How To Prepare For A Rainy Period Or A Big Storm:

This information is to help you when the severe, unexpected weather patterns of climate change occur. If you are in an area which is expected to get a lot of heavy rainfall, pay special attention to these steps well in advance. Many of these tips will also help you before the occasional sudden downpour.

Don't Be Outside In A Thunder and Lightning Storm!

Personal Concerns

☐ Check the weather forecast so you are not caught unawares.

☐ For a thunderstorm, you will want to have shelter or a safe place to wait out the storm, avoiding areas where lightning might strike outdoors.

☐ Have appropriate clothing ready, with warm sweaters

☐ Have an umbrella.

☐ Take Vitamin C to help prevent colds and flu. Eat fruits and vegetables.

☐ Be prepared for mosquitos if they come in your area.

☐ Prepare a 3-day supply of clothes, food, medication and water. One to two weeks of supplies are good to have on hand.

☐ Smaller grab-and-go kits for each member of the household, including pets, should be easy to get at in case you have to leave your home.

☐ Get a tune-up and gas for your car so it does not have problems in the rain.

☐ Review your insurance to be aware of what personal and household coverage you have.

Review Your House

☐ Move large pictures off the walls from above beds, chairs and couches.

☐ Put latches on cabinet doors.

☐ Use strong strapping or ratchets, or other connectors, to secure refrigerators, freezers, washers or dryers.

☐ Anchor water heater to the wall with straps. If necessary, get a licensed gas fitter to install a flexible gas line.

☐ Water damage on inside ceilings and walls can indicate roof leaks to be repaired.

☐ Know how to turn off your electrical panel, water and gas in case of leaks, or if directed by officials.

Outdoors

☐ Clean and repair gutters, look for loose or damaged shingles.

☐ Trim branches off trees and bushes next to the house that could strike it.

☐ Consider installing weather stripping on doors and windows.

☐ Look for places where water could collect next to the house.

☐ Sandbags could help for diverting water, debris, or mud if flooding happens.

☐ Check property hillsides for signs of erosion.

☐ Check street drains nearby; call the city if they are clogged.

Safety Steps

Stay home if possible, but make emergency plans with your family:

☐ When you hear thunder, go indoors!

☐ If you count less than 30 seconds between lightning and thunder, seek shelter immediately!

☐ You may be on your own for several days after a storm.

☐ Phone, gas, electric or water services might be disrupted. Roads could be blocked, stores and gas stations might be closed.

☐ Make emergency plans of where to go and meet the rest of your family. Think of children, pets, and those with special needs, and the rest.

☐ When the weather slows down after the storm, stay calm and give things a chance to settle.

☐ Follow directions from local authorities.

☐ If you suspect a gas leak, turn off the gas valve and leave the house.

☐ If your home is damaged, take your bags and evacuate to a safe place.

Some Articles On The Internet That Have Further Information:
National Weather Service, "Prepare! Don't Let Severe Weather Take You By Surprise"Canada, "Severe Storms – What to Do?" 8/2022
Canadian Red Cross, "Thunderstorms: Before, During & After," 2023

14. DRY WEATHER & HEAT WAVES

Here's How To Prepare For A Dry Period:

(Or To Save Water For The Future - For Your Community, Or For Sharing)

Use Water More Efficiently

☐ Jeans, t-shirts, and other items need large volumes of water to grow the cotton. Try to mend and keep your clothes longer so you don't need to buy new ones.

☐ Turn the tap off while you are brushing your teeth, shaving, bathing the dog, or washing the car, until it's actually needed for rinsing.

☐ Products such as showerheads, faucets, and toilets have water-efficiency ratings.

☐ Use the minimal amount of dishes so you don't use extra water for washing them.

☐ You can use the water you cooked vegetables or pasta in for watering plants.

☐ If you water your plants at a cool time of day, the water won't evaporate.

☐ Try not to flush the toilet more than a few times a day.

☐ If you put an air floater, or one or two empty plastic bottles with sand in them, in the tank to displace the water then each flush uses less water.

☐ A leaky toilet can waste 200 gallons of water per day. Toilet and faucet leaks should be fixed right away!

Use Less Hot Water

You can minimize the electricity needed to pump, treat, or heat water by using less water.

☐ If the water heater has a temperature control, you can adjust the output temperature to 120F to save energy.

☐ You can install water-efficient fixtures and low-flow showerheads that save hot water.

☐ Try and take fewer and shorter showers. The optimum length of a shower is considered to be five minutes by some people. Sometimes a sponge bath will be enough.

☐ Heating the water for a bath uses less electricity (fewer emissions) than a shower.

☐ Washing clothes in cold water reduces hot water usage.

Be Smart Outdoors

☐ Be smart when watering your lawn. Just water it when it's needed - early mornings are best.

☐ Use a rake or a broom instead of a water pressure device for cleaning the patio or the blacktop on your driveway.

☐ If you aerate your lawn periodically it will help rainwater soak into it and need less watering.

☐ Save rain water in a barrel and use it to water your garden, lawn, and indoor plants.

Water For Wildlife

☐ If you have wildlife in your vicinity, you can put water for them to drink in dry times at some distance from your house.

—

Some Articles On The Internet That Have Further Information:
American Red Cross, "Drought Preparedness & Water Conservation," 2023
Ready, "Drought," 5/2022

15. FOREST FIRES

To Prepare:

☐ Pay attention to community alerts, and air quality alerts.

☐ Make an Emergency Plan that everyone in your household knows about and understands if you need to evacuate quickly.

☐ Make an Emergency Plan for the office, kids' daycare, and anywhere you spend time regularly.

☐ Check that your insurance policies and ID are up to date.

☐ Use fire-resistant materials to build, renovate, or make repairs to your home.

☐ Set up a hose, from an outdoor water source, that can reach any area of your property.

☐ Create a fire-resistant zone that is free of leaves, debris or flammable materials for at least 30 feet from your home.

☐ Plan to use a room that can be closed off from outside air. Close all doors and windows.

☐ Set up a portable air cleaner to keep indoor air clean when smoky conditions exist. Use high efficiency filters in your air conditioning system to capture fine particles from smoke.

☐ You may have to evacuate quickly due to a wildfire, so learn your evacuation routes from the authorities. Practice with the members of your household, including pets.

☐ Make plans with friends or family to shelter with them where you may be safe and fairly comfortable.

Having Supplies Ready

☐ Have supplies for your household, including a first aid kit, in a to go bag or car trunk.

☐ Gradually stock up on food and essential purchases in advance.

☐ Be cautious about carrying flammable or combustible household products such as aerosols, cooking oils, and hand sanitizer.

☐ You may want to have an N95 mask ready to protect yourself from smoke inhalation.

☐ Keep your cell phone charged when wildfires could be in your area. Purchase backup charging devices to power electronics.

Safety Concerns During A Fire

☐ Pay attention to emergency alerts for information and instructions.

☐ Wet down your roof, your house and your property as much as possible if you are concerned about a fire coming close to you.

☐ Evacuate immediately if authorities tell you to do so!

☐ Check with local authorities for the latest information about public shelters.

☐ If trapped, call 9-1-1 and give your location, but be aware that emergency response could be delayed or impossible. Turn on lights to help rescuers find you.

☐ Use an N95 mask to protect yourself from smoke inhalation, or limit your exposure to smoke by going into your airtight room. Run your portable air cleaner or air conditioning filter.

☐ If you are not ordered to evacuate but smoky conditions exist, stay inside in a safe location or go to a community building where smoke levels are lower.

☐ If you are sick and need medical attention, contact your healthcare provider for further care instructions and shelter, if possible.

Returning Home After a Forest Fire

☐ Do not return home until authorities say it is safe to do so.

☐ Avoid hot ash, charred trees, smoldering debris and live embers. The ground may contain pockets of heat that can burn you or spark another fire.

☐ When cleaning, wear protective clothing during cleanup efforts — including a long-sleeved shirt, long pants, work gloves and sturdy thick-soled shoes.

☐ Wet down debris to minimize breathing dust particles. People with asthma, COPD and/or other lung conditions should take precautions in areas with poor air quality, as it can worsen symptoms.

☐ Take photographs of property damage. Make an inventory of damages and contact your insurance company for assistance.

☐ Send text messages or use social media when trying to contact family and friends. Phone systems are often busy following a disaster. Make calls only in emergencies.

—

Some Articles On The Internet That Have Further Information:

Global Forest Watch, "Forest Fires & Climate Change," 2023

Environmental Defense Fund, "Here's how climate change affects wildfires," 2023

16. PARTICIPATING IN YOUR COMMUNITY

You Might Find Projects Like These For Climate Change That You Can Participate In Near Your Home:

☐ Group organization, requesting grants and fundraising, applying to the government for assistance.

☐ Graphic design.

☐ Tree planting.

☐ Opening streams and rivers for fish.

☐ Starting a community garden.

☐ Helping wildlife.

☐ Volunteering for an organization that protects and educates people about wildlife.

☐ Joining a local club focused on advocating for climate change issues. Hosting an event for them.

☐ Sharing facts about climate change and the importance of sustainability and environmental education with your friends and on social media.

☐ Donating, or hosting a fundraiser for a local climate change or environmental non-profit organization.

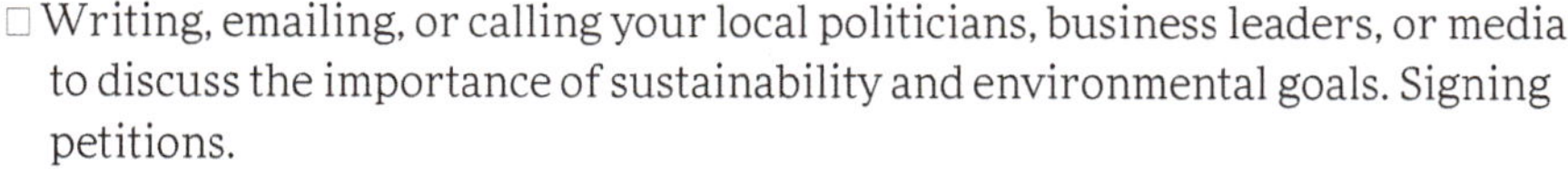

☐ Beach or community cleanup at a park, and recycling of the debris picked up.

☐ Building bike trails.

☐ Erecting alternative housing.

☐ Writing, emailing, or calling your local politicians, business leaders, or media to discuss the importance of sustainability and environmental goals. Signing petitions.

☐ We could have a volunteer database in our country with projects that have positions open in various locations.

☐ Sometimes in your community you will see brochures or handouts for projects that list things that people can volunteer for. Watch for these!

There are many other things you could do in your community as time goes on...

Some Articles On The Internet That Have Further Information:

HelpGuide, "Volunteering and its Surprising Benefits," 1999-2022

DoSomething, "6 Volunteering Ideas To Help The Environment," 4/2022

This is a comprehensive handbook
of 16 sections of how to help end Climate Change on the planet.

I have studied this for over two years
from reading many, many articles on the Internet,
and have included all the topics that seem relevant for
people to be concerned about in their everyday lives.

Also included are emergency procedures of what to do
regarding climate change weather
such as storms, droughts, and the wildfires
that sometimes start from lightning.

The book was written in Canada, and it should also
be helpful in many other parts of the world as well.

Teya France